YOUR KNOWLEDGE HAS VALUE

- We will publish your bachelor's and master's thesis, essays and papers

- Your own eBook and book - sold worldwide in all relevant shops

- Earn money with each sale

Upload your text at www.GRIN.com and publish for free

REHAN JAMIL

An Overview of Photovoltaic Power Generation and Solar PV Technology in Rural Areas of Pakistan

GRIN Verlag

Bibliografische Information der Deutschen Nationalbibliothek:

Die Deutsche Bibliothek verzeichnet diese Publikation in der Deutschen National-
bibliografie; detaillierte bibliografische Daten sind im Internet über http://dnb.d-
nb.de/ abrufbar.

Imprint:

Copyright © 2014 GRIN Verlag GmbH
Druck und Bindung: Books on Demand GmbH, Norderstedt Germany
ISBN: 978-3-656-57962-5

This book at GRIN:

http://www.grin.com/en/e-book/266832/an-overview-of-photovoltaic-power-
generation-and-solar-pv-technology-in

GRIN - Your knowledge has value

Der GRIN Verlag publiziert seit 1998 wissenschaftliche Arbeiten von Studenten, Hochschullehrern und anderen Akademikern als eBook und gedrucktes Buch. Die Verlagswebsite www.grin.com ist die ideale Plattform zur Veröffentlichung von Hausarbeiten, Abschlussarbeiten, wissenschaftlichen Aufsätzen, Dissertationen und Fachbüchern.

Visit us on the internet:

http://www.grin.com/

http://www.facebook.com/grincom

http://www.twitter.com/grin_com

An Overview of Photovoltaic Power Generation and Solar PV Technology in Rural Area of Pakistan

Rehan Jamil[1], Ming Li[2], Xu Ji[2] and Xi Luo[2]

[1]*School of Physics and Electronic Information, Yunnan Normal University, Kunming, China,*
ch.rehan.jamil@gmail.com
[2]*Solar Energy Research Institute, Yunnan Normal University, Kunming, China, lmlldy@126.com*
[2]*Solar Energy Research Institute, Yunnan Normal University, Kunming, China, jxime@126.com*
[2]*Solar Energy Research Institute, Yunnan Normal University, Kunming, China, luoxi303@126.com*

Abstract:

In this paper, we examine the solar energy which is the ultimate and free source of energy. Pakistan is situated in the sunny belt that is ideal for the solar energy technologies. This energy is abundantly available in the country as sun shines almost the whole year. The electricity in Pakistan is now facing a serious energy crisis. Despite of strong economic growths during the past decade and consequent rising demand of energy, no valuable steps have been taken to install new capacity for generation of the required energy. In this respect, solar PV is an option for electricity generation. A very large part of the rural population does not have the facility of electricity because they are either too remote or it is found extremely expensive to connect their villages to the national grid station. We can easily utilize the solar PV in some areas of Pakistan; this kind of energy is becoming familiar. Finally, in this paper importance of solar PV technology in rural area is discussed. Geographically, overview of solar radiation and solar PV option for electricity generation for the country is explored from multiple perspective comprising technical, economic, social, environmental and political aspects.

Keywords:

Solar energy, Solar radiation, Photovoltaic power generation, Photovoltaic (PV) Technology, Pakistan.

1. Introduction

In Pakistan, about 43% of the total population live without access to electricity. The 70% of this under-served population lives in rural areas, approximately 50,000 villages completely detached from the national electricity grid and most of them are below the poverty line. Many of these villages are far from the main transmission lines of the national grid, because of their relatively small population is usually not economically feasible to connect these villages to the grid [1].

In this respect, solar PV technology stands out to be one of the prospective sources to combat this adverse situation. Pakistan is in a very favourable position in respect of the utilization of solar energy. Solar energy has excellent potential in areas of Pakistan that receive high levels of solar radiation throughout the year. Every day, the country receives an average of about 19 MJ/m^2 of solar energy. However, solar energy can produce power at the point of demand in both rural and urban areas.

Corresponding author address: School of Physics & Electronic Information, Yunnan Normal University, Kunming 650092, PR China. Tel. /fax: +86 18313855339. E-mail address: ch.rehan.jamil@gmail.com (Rehan Jamil).

Solar PV electricity is an equally significant energy option for developing countries, Because of the cost of transmission lines, difficult of transporting fuel to remote areas and issues related to security of energy supply in developing countries are increasingly turning to solar energy as a cost-effective way to supply electricity.

Most governments in the developing countries give high priority to rural electrification to meet economic, socials, political and regional development goals. Several governmental and private organizations are working to install solar PV technology in rural Pakistan to meet basic energy needs. The application of PV technology in rural area is indirectly increasing the income as well as the living standard of the rural people.

Under Solar Village Electrification Program, Alternative Energy Development Board (AEDB) has been installed 3,000 Solar Home System (SHS) in 49 villages of district Tharparkar, Sindh. Another 51 villages in Sindh and 300 villages in Balochistan are approved for electrification using solar energy and will be implemented on release funds [8,14].

Another important initiative of the FATA (Federally Administered Tribal Area) Development Authority is Solar Electrification Program. Solar electricity was provided to 42 villages all over FATA and security check posts from Takhta Baig to Torkhum in Khyber Agency under the Solar Electrification Program during 2011-2012 [16]. Keeping in view the increasing demand of energy in the country, the FATA Development Authority (DA) is exploring alternate sources of energy. For this purpose, FATA DA has completed a feasibility study to provide solar energy to 450 villages of FATA. Actual electrification of about 44 villages has already started under two schemes.

2. Solar PV power generation- A global scenario

Solar PV power is the conversion of solar radiation into electricity using solar PV panels. Solar panels vary in size and power capacity, with individual panels ranging from a watt to a couple of hundred watts. These individual panels can be connected together to for a solar PV array that can be up to a megawatt in capacity. The largest single solar array in the world is installed in Utah, USA and is of 1.65 MW. Solar arrays are modular in nature and therefore solar power plants, comprising of multiple solar arrays, can be of any size from the watt- kilowatt scale to the gig watt scale (a 2,000 MW solar PV plant is under development in Tunisia, expected to be on line in 2016).

Solar PV, as an alternative to fossil fuels, is plentiful; renewable, widely distributed, cleans, and produces no greenhouse gas emissions during operation. It is widely expected that by the end of 2015 over 150 GW of solar PV power could be installed around the globe.

However, there are several limitations that need to be considered regarding solar PV power. Firstly, the technology requires a large amount of land (5 acres per MW can be considered a rule of thumb), which is a limiting factor to how much solar energy can be installed, and secondly, most renewable technologies as they are today cannot be used for base load applications due to the intermittency of energy supply caused by seasonal and time of day variations in output. This limits the share of solar energy to about 20% of the total installed capacity in any grid system as increasing it beyond this threshold could result in issues with grid stability [15].

Driven by advances in technology and increases in manufacturing scale and sophistication, the cost of solar PV panels has declined steadily since the first solar cells were manufactured and the levelized cost of electricity from PV is competitive with conventional electricity sources in an expanding list of geographic regions. Furthermore, net-metering and financial incentives, such as (Feed-in-Tariff) FIT/Upfront Tariffs for solar-generated electricity, have supported solar PV installations in many countries.

2.1. Global trends in solar PV power

Solar PV technology is the fastest growing power generation technology in the world with PV capacity being installed in over 100 countries. There has been a steady increase in the year-end capacities of solar PV capacities installed, particularly from the year 2008 onwards. Though European countries dominate the market, the technology continues to grow in Asia-Pacific, North

America, Middle East, North Africa and South Asia. The following table shows the increasing solar PV capacities globally.

Table 1. Global year-end capacity for solar PV

Photovoltaic power worldwide GW	
2005	5.4
2006	7.0
2007	9.4
2008	15.7
2009	22.9
2010	39.7
2011	67.4

3. Solar energy sector in Pakistan- Current scenario

Pakistan has an enormous potential of solar energy. Most parts of the country are blessed with very favourable conditions for development of solar energy for power generation. National Renewable Energy Laboratory (NREL), Colorado-USA, in collaboration with (United States Agency for International Development) USAID, (Pakistan Meteorological Department) PMD and AEDB, has carried out a detailed analysis to determine solar-energy potential in various regions of Pakistan and has prepared solar maps of Pakistan [17]. The Annual Global Horizontal Solar Radiation Map of Pakistan for PV power developed by NREL as given in Fig.1 below shows that most parts of Pakistan are very prolific for development of solar PV power plants. The NREL study shows that the theoretical solar energy potential for power generation in Pakistan is approximately 2.9 million megawatts [17].

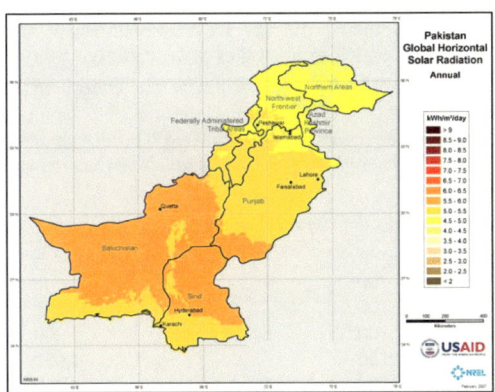

Fig.1. shows solar radiation map of Pakistan

Source: Renewable Energy Laboratory (NREL) of USA

Pakistan's geographical location is highly favourable for utilization of solar energy. It has been estimated that over the total geographical area, that is, 796,095 km^2 the available solar energy potential is 5.32 PJ/m^2/year [2]. Pakistan receives 9-10 hours sunshine in summer, 7-8 hours in winter and an average solar radiation of 5-8 kWh/m^2/day with an 85% persistence factor in more than 95% of its area [3,4,10]. The annual incident solar radiation in the country is in the range of 1900-2200 kWh/m^2 [5]. The annual mean values of insolation are from 19.0 MJ/m^2/day over most parts of the country [3,4]. The mean values of isolation are from 12-13 MJ/m^2/day in December, January to 25-26 MJ/m^2/day in June [6].

The south-western province of Balochistan and north-eastern part of Sindh offers excellent conditions for harnessing solar energy where the sun shines between 9-10 hours daily or about more than 2300-2700 hours per annual. The greatest value of 15.5 MJ/m^2/day occurs along the coastline of Balochistan and central part of Sindh in the month of January. Southern western province of Balochistan is particularly rich in solar energy with an average daily global isolation of 19-20 MJ/m^2 a day (1.93-2.03 MWh/m^2 in a year) and the annual mean sun shines duration of 8-8.5 hours. Such conditions are ideal for Photovoltaic (PV) and other solar energy applications. The Energy Information Administration describes the daily solar energy potential for Pakistan as 5.3 kWh/m^2 (1.93 MWh/m^2 annually) [7]. The monthly daily mean solar energy profile of major cities of the country is shown in Table 2.

Table 2. Monthly mean solar radiation in kWh/m^2/day in major cities of Pakistan

Month	Karachi	Lahore	Multan	Peshawar	Quetta
JAN	4.38	2.97	3.32	0.35	3.91
FEB	4.92	3.86	4.27	4.06	4.59
MAR	5.82	5.04	5.20	5.07	5.64
APR	6.23	5.79	6.30	6.02	6.63
MAY	6.44	6.32	7.70	6.96	7.61
JUN	6.30	6.18	6.52	7.19	8.22
JUL	5.31	5.70	6.42	6.50	7.45
AUG	4.94	5.20	6.02	5.91	7.32
SEP	5.50	5.20	5.56	5.37	6.70
OCT	5.38	5.06	4.64	4.40	5.64
NOV	4.66	4.19	4.84	3.53	4.48
DEC	4.12	3.44	3.20	3.20	3.77

Source: Pakistan Meteorological Department (PMD)

The Government of Pakistan (GoP) is determined to include renewable energies in the overall energy mix of the country. Alternative Energy Development Board (AEDB), created by the Government to promote and develop RE technologies in the country.

AEDB has been specifically tasked to develop renewable energy in the country for power generation through the private sector. The GoP has set an ambitious target of having at least 5% of the total power generation of the country (i.e. 9700 MW) through alternative and renewable energies (AREs) including solar energy by 2030. Pakistan's solar energy development plans see in Table 3.

Table 3. Medium term solar energy development plan 2011-2020

Year	Capacity Installed(MW)	Cumulative MW of solar Energy Installed by Year End
(2005-2010)	700	*Short term plan (2005-2010)*
2011	100	800
2012	100	900
2013	150	1,050
2014	200	1,250
2015	250	1,500
2016	250	1,750
2017	300	2,050
2018	300	2,350
2019	350	2,700
2020	300	3,000

Source: Board of Investment, Government of Pakistan

The Government of Pakistan is looking at the development of at least 500 MW solar PV power plants in the short term. AEDB has made significant progress in order to achieve this target through the integrated efforts of all the stakeholders involved in the project; the stakeholders being AEDB, Government of Sindh, (National Transmission & Despatch Company) NTDC / (Karachi Electric Supply Company) KESC, (National Electric Power Regulatory Authority) NEPRA and the Private Investors. AEDB, at its end, has done level-best to mobilize actively the stakeholders involved with the project. In this regard, AEDB has attracted national and international investors and LOIs have been issued to 7 (Seven) private investors, so far, having sound financial and technical background, for development of solar power projects.

3.1. Effects being done to meet the Gop target

AEDB is striving to give a boost to solar power generation in the country and expeditiously carrying out multidirectional activities to begin solar power projects with potential sites. The Government of Pakistan (GoP) intends to involve the private sector in these mega projects so as to ensure the sustainability of the solar energy sector. Through continuous promotion regarding solar energy potential and its prospects for utilization of power generation, AEDB has been able to attract and engage private national and international investors. Currently AEDB is working with 7 active solar PV power projects. Government of Punjab has allocated 5,000 acres land for solar power development in the Cholistan area. Private companies are being allocated required chunks of land in this area. Some companies also have acquired their own land and are developing their feasibility studies [22].

At present, around 250 MW solar PV power projects are under development. Most of the project sponsors executing these projects are conducting feasibility studies and collecting site specific data and making headway towards getting tariff from NEPRA. M/s Tech Access, M/s DACC Power Generation, M/s Buksh Energy, M/s Roshan Pakistan are the front runner companies and are in advanced stages.

Based on international experiences of solar PV, it has been deduced that the promotion of solar PV power plants can best be ensured by announcing FIT/UPFRONT TARIFF. The normal COST PLUS Tariff mechanism is not workable for these projects owing to various certain variables which make it difficult to draw a conclusive determination.

In order to promptly resolve the tariff related matters and to promptly develop the solar energy sector by harnessing the available potential and supplement its efforts to meet energy needs of the country, the GoP is looking at announcing FIT/Upfront Tariff for solar PV power projects. It is deemed that FIT/Upfront Tariff can be an effective instrument for the prompt development of solar energy sector. This has prompted AEDB to develop this mechanism and apply to NEPRA for FIT/Upfront Tariff for solar PV power projects in Pakistan.

3.2. Current development and future targets

- In Solar Energy, 6 LOIs for a cumulative capacity of 148 MW On-Grid solar PV power plants have been issued by AEDB. Additionally 3 LOIs of 70 MW capacities have been issued by Punjab Power Development Board (PPDB) [8].

- In 2013, Pakistan has planned to launch a 50MW electricity generation plant designed to produce electricity by solar energy in the Cholistan Desert in the province of Punjab, in collaboration with the German Energy Company Conergy and Hong Kong-based Ensunt. After completion, the plant will be the biggest solar power plant in Pakistan. It employs solar photovoltaic (PV) technology that converts solar energy directly into electricity, while emitting zero greenhouse gases (GHG) into the atmosphere. The project is planned to be carried out in phases of 5 MW each and would become operational and start generating electricity. The generated electricity will be supplied to the national grid. The area of this project will be about 500 Acres and annual amount of electricity going to the grid will be about 79,147 MWh. This plant will eventually supply power to more than 30,000 households in Pakistan [19]. Also, Pakistan has set a target 50 to 100 MW of photovoltaic which are expected to be installed in 2013 and at least 300 MW in 2014.

- In 2012, Pakistan has stepped ahead by inaugurating the first ever solar power on-grid power plant in Islamabad. "Introduction of Clean Energy by Solar Electricity Generation System" is a special grant aid project of the Japan International Cooperation Agency (JICA) under the Cool Earth Partnership. This project includes the installation of 178.08 kW Photovoltaic (PV) Solar

Systems each at the premises of the Planning Commission and Pakistan Engineering Council, Islamabad. This is the first on-grid solar PV project [18]. Table 4. Shows solar power project.

Table 4. Solar power project

Technology	Name/site of project	Province	Capacity (MW)
Solar PV	Cholistan	Punjab	50
Solar PV	Kasur	Punjab	10
Solar PV	Kalar Kahar	Punjab	2
Solar PV	Sanjwal (Attock)	Punjab	1
Solar PV	Pind Dadan Khan	Punjab	10
Solar Thermal	Lodhran	Punjab	20

Sources: Board of Investment, Government of Pakistan

4. Solar PV technology in rural area

According to the economic survey report 2011-2012, under Solar Village Electrification Program 3,000 Solar Home Systems (SHS) have been installed in 49 villages of district Tharparkar, Sindh. Another 51 villages in Sindh and 300 villages in Balochistan have been approved for electrification using solar energy and the same will be built following release of funds [8,14].

Each household in each village has been provided with 120 Watt/ 80 Watt/ 40 Watt Solar Panels, Charge Controller, Battery, 4 CFL Lamps, 2 LED lights, a 12 Volt DC fan and a TV socket. The solar panel charges the battery during daylight hours, and the stored energy is used to run the electrical appliances throughout the day and night. The user is only required to switch the lighting system on/off, as is done in regular home lighting systems. In addition, a Solar Disinfecting Unit and a Solar Cooker have also been provided to each household [8,9,14]. Table 5,6,7. Below shows villages electrified using solar energy in different provinces.

Table 5. Villages electrified through Solar Photovoltaic during 2004-2005

Name of Village	District	Province	N0. of Houses
Narian	Rawalpindi	Punjab	100
Khorian	Rawalpindi	Punjab	135
Allah Baksh Bazar Dandar	Turbat	Balochistan	121
Qadirabad	Turbet	Balochistan	53
Lakhi Bhair	D.G khan	Punjab	115
Bharomal	Chachro	Sindh	115
Jhanak	Kohat	K.P.K	120
		Total	**759**

Source: Board of Investment, Government of Pakistan

Table 6. Village electrified through Solar Photovoltaic during 2005-2006

Name of Village	District	Province	N0. of Houses
Khirzaan	Khuzdar	Balochistan	100
Basti Bugha	D.G Khan	Punjab	100
Pinpario	Chachro	Sindh	100
Shnow Garri	Kohat	K.P.K	100
Takht, Mamamacharzai	Kallat	Balochistan	100
Kili Mama Macherzai	Killa Saifullah	Balochistan	100
		Total	**600**

Source: Board of Investment, Government of Pakistan

Table 7. Village electrified through Solar Photovoltaic during 2007-2012

Name of Village	District	Province	NO. of Houses
Mithi	Tharparkar	Sindh	999
Diplo	Tharaparkar	Sindh	454
Chachro	Tharparkar	Sindh	1,252
Nangarparkar	Tharparkar	Sindh	391
		Total	**3,096**

Source: Alternative Energy Development Board (AEDB)

Under the Parliamentarian Sponsored Village Electrification Program (PSVEP), 119 Solar Home Systems have been provided to 10 villages of Deh Tiko Baran District Jamshoro, Sindh and 200 Solar Home Systems in 16 Villages of Karak, District Khuzdar, and Balochistan [8]. Table 8. shows villages electrified under Parliamentarian Sponsored Village Electrification Program.

Table 8. Parliamentarian Sponsored Village Electrification Program, Sindh and Balochistan

Name of Village	District	Province	NO. of Houses
Deh Tiko Baran	Tharparkar	Sindh	119
Karak	Khuzdar	Balochistan	200
Baiker	Dera Bugti	Balochistan	100
		Total	**419**

Source: Alternative Energy Development Board (AEDB)

Based on the success of the Solar Village Electrification program, the Government of Pakistan had approved replication of this project in 400 villages in Balochistan and Sindh. The overall objective of this program is to develop technological and strategic implementation plan to electrify 6968 villages in Balochistan and 906 villages in Sindh through Renewable Energy sources that are outside the 20 km radius of the national grid [8,14]. It is planned to implement this program with a phase wise and clustered approach. In these projects 100 villages selected in Sindh province and 100 villages each in Balochistan Province i.e. Southern Balochistan, Central Balochistan and Northern Balochistan will be electrified through SHS technology.

The project is currently in the implementation phase and the transportation / installation of the equipment in the homes of the selected villages is underway. People inhabiting these remote villages will have access to not only the basic amenity of electricity, but clean potable solar pumped water as well, thus elevating their standard of living.

4.1. Solar electrification project at Tharparkar

Eventually, the renewable energy has come to the rescue of Thar. AEDB launched project for 3,000 homes in Taluka Mithi, Diplo, Chachro, and Nagarparkar. The object of the rural electrification project, designed and launched by AEDB, was to provide electricity for lighting, means of communication and clean drinking water facilities to the villages where it is technically and economically not feasible to extend gird connectivity. Under this project, 906 villages have been identified in a province to be electrified through alternative energy resources by AEDB in phase I [12-14]. In this respect, it is blessing for the Tharparkar district for having a major portion of village electrification in a single district.

There are other remote areas which are also not connected with national gird such as coastal belt of Badin, Thatta, the remote areas of Manchar Lake, entire Kohistan belt, and areas of "Achhro Thar" at Sangher. All these areas are also supposed to be provided reasonable proportion as those are also under the same condition.

Rural Electrification Project (REP) Mithi office was established in June 2007 and is heavily engaged to electrify the backward areas of" Thar "desert through Solar Home System (SHS). A team of trained skilled manpower including engineers and other staff has successfully electrified 50 villages of Tharparkar district. Fig.2 shows an SHS installed by AEDB in a village of Tharparkar, Sindh. Remarkable socioeconomic and sociocultural change has been visualized in the dark of Tharparkar which have been brightened through solar energy [14].

Fig.2. shows a rural electrification project in Sindh (by author)

Beside rural electrification through SHS, AEDB is also working on solar water pumping and desalination, because water is the basic need of human being and ground water, i.e., dug well and rainy pounds is the only source of the almost whole population of the district Tharparkar. Solar water pumping along with desalination well certainly helps to improve their rate of production of handmade things so the economic condition of the poor Thar peoples will also be improved. Fig.3 shows successfully Solar Home System (SHS) installed in Tharparkar.

Fig.3. a view of Solar Home System installed in Tharparkar (by author)

4.2. Solar energy to village in FATA

FATA DA's Solar Electrification Program is the first ever alternate energy initiative in the region. The program started in 2008 has been appreciated by all the stakeholders including Parliamentarians, Political Administration and the locals. Till June, 2012, 42 villages have been electrified. Under this program, Solar Home Systems, Solar Pumping Systems and Solar Street Lights are covered [16]. Keeping in view the energy shortage, the program is being expanded. Fig.4 shows solar panels in FATA.

Fig.4. shows Solar Panel at Torkham, FATA (by author)

As this was a new initiative, some deficiencies were noticed in the initial phase. The Authority has now amended the strategy under which major health facilities, offices of the Political Administration, public places and bazaars like Bab-e-Khyber would also be provided solar energy in addition to the private sector [16]. All stakeholders will be involved to maximize its benefits. Fig.5 a view of Solar Street Light (SSL) at Torkham Bazar.

Fig.5. a view of Solar Street Light at Torkham Bazar, FATA (by author)

4.2.1. Provision of solar energy to village in FATA

The demand of Energy has increased tremendously during the last few decades in Pakistan; the same is expected to increase further in the coming years. Solar energy is the possible clean and low cost renewable resources available in the country. The use of such natural available energy resources can suitably be utilized in the remote areas of FATA where the supply and maintenance of electric supply from the national grid is very expensive.

The scattered and distantly located villages of FATA are either deprived of electricity or the power supply available is uncertain and with fluctuating/low voltage. Like other parts of the country, the exploitation of alternate energy resources also carries primary importance in FATA. Detailed feasibility study for provision of Solar Energy System in 450 villages of FATA has been awarded to NUST (National University of Science and Technology) Consulting firms. Based on the recommendations of the study the PC-I will be framed to cover the cost of actual implementation of solar electrification and solar pumping of drinking water in 450 villages of FATA.

4.2.1.1. Project scope

Under this project solar electrification systems will be installed in 450 selected villages of seven Agencies and six Frontier Regions of FATA. Besides, solar pumping for drinking water will also be installed in 250 villages. Solar power systems will be provided in remote villages where presently there is no electricity supply, and there is also no possibility of providing electricity from the national grid in the near future.

4.2.1.2. Benefits

The project will provide solar electricity to 450 villages of FATA besides provision of solar pumping of drinking water supplies to 250 such villages. This project will have a visible positive effect on the social status of the deprived people of these remote areas.

4.2.1.3. Present status

Provision of solar energy to villages in FATA has been taken up by FATA DA. NUST Consulting had surveyed 450 villages. A feasibility study of 450 villages had been completed, PC-1 (Phase-I) of 22 villages was approved amounting to Rs.195.98 million.

PC-1 (Phase-II) for 16 villages has been approved amounting to Rs.171.15 million. Due to the unavailability of security in some villages, the systems were diverted to alternate villages.

As a result, number of Phase-II approved villages came to twenty four (24) villages. Tender for procurement and installation for the (Solar Home System) SHS, (Solar Pumping System) SPS and (Solar Street Light) SSL was awarded to national contractor M/S Solar Tech Lahore.

M/S Solar Tech completed procurement and installation of Solar systems in 22 villages for Phase-I and in 24 villages for Phase-II. Table 9 number of various systems to be installed through this project is as below.

Table 9. Number of various systems to be installed through this project

Project	SHS	SSL	SPS	Solar geyser	Solar cooker	Solar tube wells
Phase-I	1055	28	3			
Phase-II	1022	14				
Phase-III				500	500	12
Phase-IV	1346	37	14			

Sources: FATA Development Authority

(Phase-III) and (Phase-IV), apart from villages, Solar Systems will also be installed in Government hospitals, press clubs and PA (Political Administration) Compounds.

Conclusion

This paper addresses Pakistan as the most potential hub in the region to generate electricity exploiting solar energy as it has immense capacity and potential to explore sunlight as compared with different part of the world. Pakistan's geographical location presents excellent conditions for deployment of small and large scale solar PV power projects. This vast potential can be exploited to produce electricity by solar PV technology, which could be provided to off-grid communities. Photovoltaic technology is one of the most reliable and efficient off-grid solution to get electricity in remote places. Due to increasing demand and consecutive industrial development, the photovoltaic technology has grown today to maturity and is ready for a number of applications. As Pakistan is facing acute energy crisis, so this technology can contribute to overcoming the hard situation and can bring sustainable development.

References

[1] T. Muneer, M. Asif, Prospects for secure and sustainable electricity supply in Pakistan, Renewable and Sustainable Energy Reviews, 2007, 11 (4): 655-671

[2] Ghaffar, Muhammad A, "The Energy supply situation in the Rural Sector of Pakistan and The Potential of Renewable Energy Technologies". "Renewable Energy", 1995, 6 (8): 941-976

[3] Ghayur Adeel, Role of Satellite for Renewable Energy Generation Technologies in Urban Regional and Urban Settings. Institute of Electrical Engineers (IEEE), 2006.

[4] Abro, L.A.R.A.RS., Solar and Wind Energy Potential and Utilization in Pakistan, "Renewable Energy", 1994, 5 (1): 583-586

[5] Asif M, Sustainable Energy Options for Pakistan, Renewable and Sustainable Energy Reviews: 2009, 13 (4): 903-909

[6] Sukhera M.B, M. B. Pasha and M.A.R, Solar radiation Maps for Pakistan, "Solar & Wind Technology", 1987, 4 (2): 224-238

[7] Sadiq Ali Shah, Yang Zhang Prospects of Coastal Solarisation for Freshwater and Electricity Production, ISESCO Journal of Science and Technology Vision, 2010, 6 (10): 82-87

[8] Pakistan Economic Survey 2011–2012. Economic Advisers Wing, Ministry of Finance, Government of Pakistan; June 2012.

[9] Pakistan Economic Survey, 2006–2007, Government of Pakistan.

[10] Munawar A. Sheikh, Energy and renewable energy scenario of Pakistan, Renewable and Sustainable Energy Reviews, 2010, 14 (1): 354-363

[11]   Rural Electrification Through PV and Hybrid Systems in Pakistan; M. N. Zakir, K. Khetran, I. A. Qazi and P. Akhter, International Meeting on Renewable Energy Technologies and Sustainable Development, COMSATS (2004) Islamabad.

[12]   Energy and Environment and Sustainable Development, Mohammad Aslam Uqaili, Khanji Harijan editors. Practical Application of Solar Energy at Desert of Tharparkar, Pakistan. Springer Wien New York Publisher. 2012, pp. 144-146

[13]   Memon MD, Harijan K, Uqaili MA (2007) Potential of solar home system in Pakistan. Proceedings, international conference on Engineering Technology (ICET 2007), University of Kuala Lumpur, Malaysia, December 11-13, 2007.

[14]   Rural Electrification Programme. Alternative Energy Development Board, Ministry of Water and Power, Government of Pakistan. Available at http://www.aedb.org/rep.htm [accessed 16.2.2013].

[15]   Sadiq Ali Shah and Rodger Edwards, the Sustainable energy generation processes in the deserts of solar-rich countries, International Journal of Sustainable Energy, vol. ahead-of-p, no. ahead-of-p, 2011, pp. 1-8.

[16]   FATA   Development   Authority   Annual   Report   2011-2012   Available at http://www.fatada.gov.pk/index.php [accessed 25.4.2013].

[17]   National Renewable Energy Laboratory (NREL), Pakistan Resource Maps and Toolkit http://www.nrel.gov/international/ra_pakistan.html.

[18]   Japan International Cooperation Agency, Pakistan gets first on-grid solar power station. Available at: http://www.jica.go.jp/pakistan/english/office/topics/press120529.html [accessed 15.2.2013].

[19]   Clean Development Mechanism (CDM), 50 MW Solar PV Power Project in Cholistan, Pakistan.   Available   at
    : https://cdm.unfccc.int/Projects/Validation/DB/UVU08XN11ZIDAN4TSFGKSQJKJHV53D/view.html [accessed 20.2.2013].